FLORA OF TROPICAL EAST AFRICA

LINACEAE

Doreen L. Smith

Trees, shrubs, lianes or herbs. Leaves simple, alternate or opposite; stipules present, rarely absent, divided, entire or gland-like. Inflorescence a terminal or axillary cyme, rarely flowers solitary. Flowers regular, hermaphrodite, hypogynous. Sepals 4–5, imbricate, free or partially united. Petals 4–5, contorted in bud, free or partially united at base, often clawed. Stamens 1–3 times as many as sepals; filaments united at base; staminodes sometimes present. Ovary superior, 2–5-locular; each loculus often subdivided by a false septum. Ovules pendulous, 2 per loculus. Styles 2–5, free or united at base. Fruit a capsule or drupe. Seeds with or without endosperm; embryo straight or slightly curved.

The limits of this family have been debated by several authors. Those observed here correspond with *Linaceae* subfam. *Linoïdeae* of Winkler in E. & P. Pf., ed. 2, 19a: 107 (1931) (excluding the tribe *Nectaropetaleae*, which is now often placed in *Erythroxylaceae*), and *Linaceae* of Hutch., Fam. of Fl. Pl., ed. 2, 1: 260 (1959), and Robson in F.Z. 2: 91 (1963).

Trees, shrubs or lianes; stamens 10, all fertile . . . 1. **Hugonia**

Herbs, sometimes woody at the base; fertile stamens 4–5:

   Sepals entire or with a glandular-serrate margin, free to base . . . . 2. **Linum**

   Sepals divided into (2–)3(–4) coarse teeth at apex, united at base . . . . 3. **Radiola**

1. HUGONIA

L., Sp. Pl.: 675 (1753) & Gen. Pl., ed. 5: 305 (1754)

Trees, shrubs, climbers or lianes with simple, appressed or spreading hairs on stems, leaves and often on floral parts. Some lateral stems short and modified as coiled hooks for scrambling. Leaves simple, alternate, petiolate; margin entire, crenulate, or serrulate to serrate. Stipules palmatifid, digitately laciniate or pinnatilobed, early deciduous or persistent. Inflorescence cymose, terminal or axillary, or flowers solitary or fascicled, axillary. Bracts similar in shape to stipules. Sepals 5, entire, unequal, rarely subequal, free, persistent. Petals 5, yellow, orange-yellow or rarely white, free, clawed. Stamens 5 + 5; filaments 5 short, 5 long, united at base forming a cup round the ovary. Ovary 2–5-locular; often with sterile and fertile loculi alternating; styles 2–5, free, or united at base; stigmas capitate. Fruit a drupe, with hard woody endocarp enclosing 2–5 flattened seeds. Embryo straight or curved, endosperm present.

A genus of about 40 species from tropical Africa, Madagascar and Mascarene Is., Indo-Malesia and New Caledonia.

Distinctive features in this genus are the coiled hooks for climbing and the remarkable divided stipules.

In the following descriptions the term peduncle is used for the flower-stalk in those species with solitary or fascicled axillary flowers, although morphologically it may be homologous with the pedicel in a cymose inflorescence.

Flowers in axillary several-flowered cymes; petals
    pubescent on both surfaces; flowering sepals very
    sparsely pubescent on outer surface, very unequal
    in shape, the outer broadly ovate to suborbicular,
    the inner emarginate and rounded, apiculate   .   1. *H. platysepala*
Flowers axillary, usually solitary, but if several flowers
    in one leaf-axil then petals glabrous except for
    apical margin; flowering sepals densely pubescent
    on outer surface, slightly unequal, acute or obtuse:
Sepals* lanceolate, acuminate; young twigs usually
    villous with long spreading hairs only; axillary
    hooks always definitely alternate   .   .   .   2. *H. villosa*
Sepals* always rounded or obtuse; indumentum of
    young twigs usually short and of spreading and
    appressed hairs; axillary hooks usually sub-
    opposite:
Sepals 9–12 mm. long; indumentum chocolate-
    brown; leaves with 9–12 pairs of lateral
    veins; tertiary venation reticulate and promi-
    nent on both surfaces   .   .   .   .   4. *H. grandiflora*
Sepals 5–9 mm. long; indumentum light brown,
    golden-brown, greenish, or silvery-grey; leaves
    with usually 12–22 lateral veins, tertiary vena-
    tion rarely prominent:
Stipules persistent, usually still present on
    second-year twigs, always digitate-laciniate,
    5–15 mm. long, with brown indumentum
    and spreading lobes.   .   .   .   .   3. *H. castaneifolia*
Stipules very early deciduous but occasionally,
    if pinnatilobed, persistent for almost a year,
    varying in shape from digitate-laciniate,
    with connivent lobes and 3–5 mm. long, to
    pinnatilobed, with spreading lobes and
    5–8 mm. long; indumentum light golden-
    brown to silvery-grey   .   .   .   .   5. *H. busseana*

1. **H. platysepala** *Oliv.* in F.T.A. 1: 272 (1868); De Wild., Pl. Bequaert.
4: 277 (1927); T.T.C.L.: 266 (1949); C.F.A. 1: 245 (1951); Wilczek in
F.C.B. 7: 43, t. 6 (1958); F.W.T.A., ed. 2, 1: 359 (1958); Lind & Tallantire,
Fl. Pl. Uganda: 62 (1962). Types: Principe I. & Fernando Po, *Mann* (both
K, syn. !) & Angola, Golungo Alto, Serra de Alta Queta, bank of R. Cuango,
Welwitsch 1584 (LISU, syn., K, isosyn. !)

Small tree, shrub or climber, 2–20 m. high. Young stems with sparse
short appressed light-brown hairs, soon glabrescent and dark brown,
becoming paler and longitudinally ridged with age. Hooks opposite,
subopposite or alternate. Leaf-blade elliptic to oblanceolate, 5·5–23 cm.
long, 3·5–8 cm. broad, shortly acuminate, undulate, slightly serrate or
serrate, chartaceous, glabrous except on 12–20 pairs of lateral veins and the
midrib which are usually appressed-pubescent. Stipules very soon deciduous,
present only at growing point of stem, digitate-laciniate, with broad basal
part and short subulate lobes, and with sparse short appressed pale
brown hairs. Inflorescence cymose, axillary, 2–14–many-flowered. Sepals
unequal; outer three very broadly obovate to orbicular, inner two emarginate, apiculate and slightly keeled, 5–7 mm. long, with both surfaces very

---

* Unless otherwise stated key descriptions refer to the outer sepals and their outer
surfaces. The inner sepals can be very variable in some species.

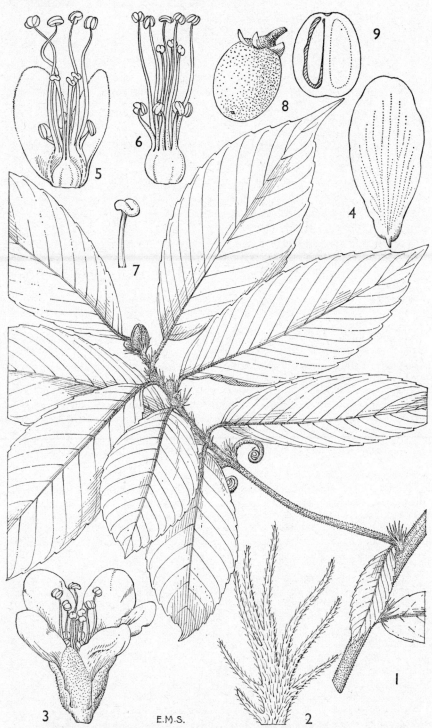

E.M.S.

FIG. 1. *HUGONIA CASTANEIFOLIA*—**1**, habit, × 1; **2**, stipule, × 4; **3**, flower, × 3; **4**, petal, × 4; **5**, section of flower, petals removed, × 4; **6**, relative position and lengths of stamens and styles, × 4; **7**, upper part of style showing grooved, capitate stigma, × 12; **8**, fruit and persistent calyx, × 1½; **9**, longitudinal section of fruit showing pendulous ovule, × 1½. 1, 8, 9, from *Drummond & Hemsley* 3573; 2, from *Faulkner* 1768; 3, from *R. M. Graham* 1707; 4–7 from *Greenway* 9638.

sparsely and coarsely pubescent. Petals obovate, 1·5-2 cm. long, shortly clawed, truncate, with short or long silky hairs on both surfaces, yellow or white. Stamens 10. Styles 4-5, free, with shaggy hairs. Drupe globose, ± 1-2 cm. in diameter, glabrous, orange, drying straw-coloured or black. Fig. 2/1.

UGANDA. Bunyoro District: Budongo Forest, Apr. 1938, *Eggeling* 3619!; Masaka District: Sese Is., Bukasa I., 27 Feb. 1945, *Greenway & Thomas* 7200!; Mengo District: Kyagwe County, Nakiza Forest, 24 Jan. 1951, *Dawkins* 703!
TANGANYIKA. Bukoba District: Kashenye, Nyakashanje R., Sept. 1957, *Salim* 9! & Rubare Forest Reserve, Sept. 1958, *Procter* 1025!
DISTR. U2, 4; T1; widespread in West Africa from Sierra Leone to Cameroun Republic; also in the Congo and Sudan Republics and Angola.
HAB. Riverine forest or secondary growth; 1050-1220 m.

2. **H. villosa** *Engl.* in E.J. 32: 105 (1902); De Wild., Pl. Bequaert. 4: 288 (1927); Exell & Mendonça in C.F.A. 1: 244 (1937); Wilczek in F.C.B. 7: 49 (1958). Types: Angola, Malange, *Marques* 299 (COI, syn.) & Congo Republic, Stanleypool, *Laurent* (BR, syn.)

Small tree, scandent shrub or liane, 6-20 m. high. Young twigs with long spreading brown villous indumentum; older stems glabrescent, light or dark brown. Hooks alternate. Leaf-blade elliptic to oblanceolate, 5-18 cm. long, 2-6 cm. wide, acuminate, shallowly serrate, both surfaces often pubescent when young, glabrescent, with midrib and the 15-21 pairs of lateral veins only remaining appressed-pubescent. Stipules laciniate with spreading subulate lobes, usually persistent for a time, 15-20 mm. long, with spreading brown indumentum. Flowers one to several but rarely more than three or four, in leaf-axils; pedicels 1-6 mm. long, stout; bracts large, shaped as stipules, up to 15 mm. long. Sepals subequal, lanceolate, 8-15 mm. long, acuminate, shortly and densely pubescent externally; inner surface glabrous and striate. Petals oblong to spathulate, 15-18 mm. long, with fleshy base and claw, ciliate at tip, otherwise glabrous, golden-yellow. Styles 5, free, glabrous. Drupe globose or ovoid, ± 1·2 cm. in diameter, orange or yellow, drying black or yellowish. Fig. 2/2.

TANGANYIKA. Kigoma District: Kabogo Mt., 21 May 1963, *Azuma* 508!
DISTR. T4; Congo Republic; Angola.
HAB. Riverine forest; ± 800 m.

VARIATION. The indumentum of young twigs varies from dense to a few long hairs only.

NOTE. This species is easily distinguished from others in East Africa because of the lanceolate-acuminate sepals. These are especially noticeable in the buds, which are conical or flask-shaped instead of globose or ovoid as in the other species.

3. **H. castaneifolia** *Engl.* in E.J. 40: 45 (1907); K.T.S.: 245 (1961). Type: Kenya, without precise locality, *Elliott* 114 (K, syn.!)

Small tree, scandent shrub or liane, 2-7 m. high. Young branches with short dense spreading and appressed brown pubescence; older stems often remaining densely pubescent for some time, but usually glabrescent and light brown or greyish-cream, with small round prominent lenticels. Hooks opposite or subopposite. Leaf-blade ovate, elliptic-obovate or elliptic, 5·5-22·4 cm. long, 2·8-8 cm. broad, from nearly entire to serrate, either softly and densely pubescent on both, or lower surfaces only, or practically glabrous when mature; lateral veins 11-22 pairs, always appressed-pubescent at least on lower surface of leaves. Stipules persistent, usually still present on second-year twigs, 5-15 mm. long, always digitate-laciniate, with brown indumentum and spreading subulate lobes. Flowers pedunculate, usually solitary but occasionally 2-4 in one leaf-axil, sweetly scented; bracts small, shaped as stipules. Sepals ovate, elliptic or oblong, obtuse, with short dense and appressed golden-brown to greenish-grey indumentum, glabres-

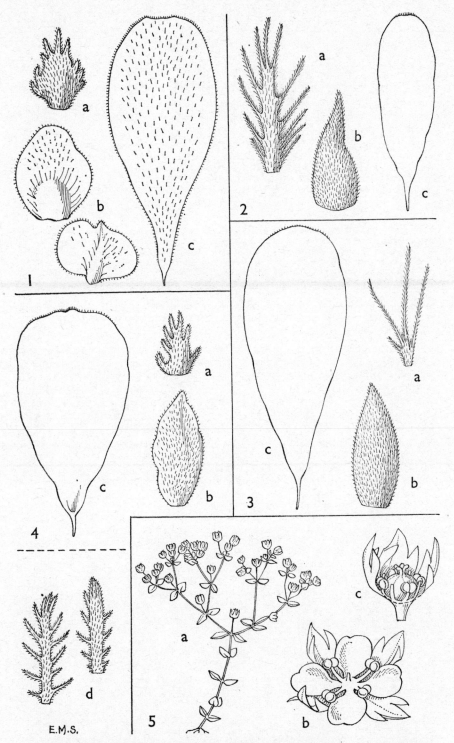

FIG. 2. *HUGONIA*—**a** & **d**, stipules; **b**, sepals; **c**, petals, of 4 *Hugonia* spp., all × 3; **1**, *H. platysepala*; **2**, *H. villosa*; **3**, *H. grandiflora*; **4**, *H. busseana*; *RADIOLA LINOIDES*—**5 a**, habit, × 2; **5 b**, opened flower, with gynoecium removed, to show united petal and sepal bases; note absence of staminodes, × 30; **5 c**, sepals, androecium and gynoecium; note presence of staminodes, × 30. 1, from *Salim 9*; 2, from *Gossweiler 13842*; 3, from *Semsei 709*; 4 a–c, from *Eggeling 6380*; 4 d, from *Milne-Redhead & Taylor 7564*; 5, from *Richards* (Tanganyika).

cent in fruit. Petals oblanceolate to narrowly obovate, 15–20 mm. long, rounded and ciliate at tip, otherwise glabrous, shortly clawed, yellow or white. Styles 2–4, glabrous, free or fused at base. Drupe obovoid, 1·2–1·5 cm. in diameter, orange, drying straw-coloured or black. Fig. 1, p. 3.

KENYA. Kwale District: Jadini, 7 Dec. 1959, *Greenway* 9638!; Kilifi District: Arabuko Forest Reserve, 17 Oct. 1962, *Greenway* 10839! & Sokoke Forest, 28 Aug. 1945, *Jeffery* K300!

TANGANYIKA. Tanga District: Pongwe-Kange forest area, 8 Jan. 1958, *Faulkner* 2116!; Pangani District: Msubugwe Forest, 5 Mar. 1963, *Mgaza* 536!; Ulanga District: near Ifakara, Funge, June 1959, *Haerdi* 253 A/O!

ZANZIBAR. Zanzibar I., Pangajuu, 10 June 1930, *Vaughan* 1336!

DISTR. K7; T3, 6; Z; not known elsewhere.

HAB. Dry evergreen or riverine forest, persisting in secondary bushland; 7–150 m.

SYN. *H. holtzii* Engl. in E.J. 40: 47, fig. 1 (1907); T.T.C.L.: 266 (1949). Type: Tanganyika, Sachsenwald near Dar es Salaam, *Engler* 3237 (B, holo. †)

VARIATION. This species varies considerably in the amount of indumentum on the leaves and young twigs.

   Semsei 1743 from Tanganyika, Uzaramo District, Kiserawe, Pugu Forest Reserve, has abnormally small, up to 5·5 cm. long leaves with only 7–9 lateral pairs of veins. The flowers are small and practically sessile. This may be a variety of *H. castaneifolia* Engl.

4. **H. grandiflora** *N. Robson* in Bol. Soc. Brot., sér. 2, 36: 6 (1962) & in F.Z. 2: 92 (1963). Type: Mozambique, Macondes, Mueda, *Mendonça* 945 (LISC, holo.)

Shrub or liane. Young twigs with short chocolate-brown spreading and appressed pubescence, becoming lighter brown or cream and glabrescent with age, often with raised lenticels. Hooks opposite. Leaf-blade oblong-elliptic to narrowly obovate, 6·5–12 cm. long, 3–5 cm. wide, shortly acuminate, remotely and slightly serrate, subcoriaceous, glabrous except on 9–12 lateral veins; tertiary venation reticulate and prominent. Stipules persistent, digitate-laciniate, with spreading subulate lobes, 5–10 mm. long, and brown indumentum. Flowers 1–2 in each leaf-axil; peduncle stout, 4–8 mm. long. Sepals ovate or elliptic, 9–10 mm. long, obtuse or subacute, with appressed chocolate-brown pubescence externally, glabrous inside. Petals obovate, 15–30 mm. long, rounded or emarginate, glabrous except for apical fringe, yellow. Styles 3–5, free, glabrous. Drupe subglobose, 1·5 cm. in diameter, brown. Fig. 2/3, p. 5.

TANGANYIKA. Lindi District: Rondo Plateau, Mchinjiri, Mar. 1952, *Semsei* 709!

DISTR. T8; also Mozambique (Macondes in Niassa Province)

HAB. Uncertain, probably dry evergreen forest; ± 500 m.

NOTE. *H. grandiflora* differs from the related species *H. castaneifolia* chiefly by the larger flowers and the leaves which have fewer pairs of lateral veins.

5. **H. busseana** *Engl.* in E.J. 40: 45 (1907) & V.E. 3(1): 722 (1915), as '*bussei*'; T.T.C.L.: 266 (1949); F.Z. 2: 93, t. 9/A (1963). Type: Tanganyika, Songea District, NE. of Songea near Madjanga, *Busse* 761 (B, holo.†, K, iso.!)

Small tree, shrub or climber; trunk with rough greyish bark. Young twigs with dense golden or light-brown short spreading pubescence, becoming paler brown or greyish-cream and glabrous with age, either corky and grooved or thin and covered with many raised round lenticels. Hooks opposite or subopposite. Leaf-blade elliptic, obovate or broadly oblanceolate, 5–14 cm. long, 2–5·5 cm. broad, acute, mucronate or rounded, entire to serrate, either practically glabrous and chartaceous, or softly and densely pubescent especially on the veins beneath; lateral veins 11–14 pairs, prominent beneath. Stipules either digitate-laciniate, 3–5 mm. long, with short connivent or spreading lobes, usually very early deciduous and with golden-brown spreading indumentum, or else pinnatilobed, 5–8 mm. long, may be persistent on first-year twigs, with appressed silvery-grey hairs. Flowers usually solitary in leaf-axils; peduncle stiff and slender, 5–25 mm. long; bracts small, shaped as stipules. Sepals ovate to elliptic or oblong, 6–9 mm. long, densely pubescent outside, practically glabrous within, greyish or yellowish-green. Petals obovate, clawed, somewhat rounded and ciliate at apex, otherwise glabrous, yellow. Styles 2–4, free, glabrous. Drupes obovoid, up to 3 cm. long, orange-yellow. Fig. 2/4, p. 5.

TANGANYIKA. Songea District: Gumbiro, 24 Jan. 1956, *Milne-Redhead & Taylor* 8505!; Tunduru District: near Tunduru on Masasi road, Nov. 1951, *Eggeling* 6380!; Lindi District, 9·5 km. S. of R. Mbemkuru crossing on Kilwa–Lindi road, 6 Dec. 1955, *Milne-Redhead & Taylor* 7564!
DISTR. T8; Zambia, Malawi, Mozambique
HAB. Deciduous woodland and dry evergreen forest; 130–910 m.

SYN. *H. buchananii* De Wild., Pl. Bequaert. 4: 269 (1927). Type: Malawi, *Buchanan* 367 (BM, holo., E, iso.)
     H. arborescens Mildbr. in N.B.G.B. 12: 513 (1935); T.T.C.L.: 266 (1949). Type: Tanganyika, Lindi District, Lake Lutamba, *Schlieben* 5188 (B, holo. †)
     H. arborescens var. *schliebenii* Mildbr. in N.B.G.B. 12: 514 (1935); T.T.C.L.: 266 (1949). Type: Tanganyika, Lindi District, Lake Lutamba, *Schlieben* 5206 (B, holo, †)
     H. faulknerae Meikle in K.B. 5: 338, fig. 1 (1950). Type: Mozambique, Quelimane District, *Faulkner* K80 (K, holo.!)

VARIATION. *H. busseana* is a very variable species, especially in amount of indumentum, size of floral parts, stipule-shape and persistence, and bark of twigs. However, the differences are not constant and intergrade from one plant to another.

2. LINUM

L., Sp. Pl.: 277 (1753) & Gen. Pl., ed. 5: 135 (1754)

Perennial or annual herbs, often woody at the base. Leaves sessile, alternate, opposite or whorled, entire or with shortly denticulate margin. Stipules glandular, persistent, or absent. Inflorescence a terminal monochasial or dichasial cyme, or rarely flowers solitary. Sepals 4–5, entire or with capitate-glandular serrate margin, free, persistent. Petals 4–5, white, yellow, blue, pink or red, free or very rarely united at base, shortly clawed. Stamens 4–5, alternating with petals; staminodes occasionally present, 4–5, filiform and between the stamens, all joined at base forming a short staminal tube. Ovary 4–5-locular; each loculus partly divided by a false septum and containing two pendulous ovules. Styles 4–5, free or rarely united at base; stigmas linear, oblong or capitate. Fruit a 4–5-locular, 8–10-valved capsule. Seeds smooth and flat, with little or no endosperm present; testa becoming mucilaginous on wetting; embryo straight.

About 230 species, mainly in north temperate regions, especially abundant in North America and the Mediterranean area; also in the tropics at high altitudes.

Capsule up to 10 mm. in diameter; flowers blue or
    white; leaves alternate . . . . . 1. *L. usitatissimum*
Capsule up to 3 mm. in diameter; flowers yellow or
    rarely white; leaves alternate, opposite or
    whorled:
  Leaves seldom in whorls, usually alternate or the
    lower ones opposite, acute:
    Leaves all alternate; sepals with midrib promi-
     nent in basal half or two-thirds . . . 2. *L. volkensii*
    Leaves, at least the lower ones, opposite; sepals
     with midrib prominent to the apex . . 3. *L. thunbergii*
  Leaves normally in whorls of three or four, ovate or
    broadly elliptic, with rounded apex . . 4. *L. keniense*

1. **L. usitatissimum** *L.*, Sp. Pl.: 277 (1753); Hegi, Ill. Fl. Mittel-Europa 5(1): 20, fig. 1676, t. 175/1 (1924); C.F.A. 1: 242 (1951). Type: N. Africa, *Herb. Linnaeus* 396.1 (LINN, syn.)

Annual herb, sometimes woody at base, up to 60–65 cm. high. Stem terete, faintly longitudinally ridged. Leaves sessile, alternate, linear-lanceolate, 30–60 mm. long, glabrous. Inflorescence a corymbose cyme; pedicels 20–30 mm. long. Sepals 5, ovate-acuminate, about 5 × 2·5 mm. in flower, enlarging to about 7 × 4 mm. in fruit, with papery translucent entire margin; main vein prominent only in basal half of sepal. Petals 5, free, obovate, 8 × 3·5 mm., narrowing to a clawed base, blue or white. Stamens 5. Ovary 5-locular. Capsule 10-valved, globose, 8–10 mm. in diameter, exceeding calyx in length, honey-brown. Seeds shining, light brown, ± 5 mm. long.

KENYA. Near Nairobi, Aug. 1903, *Whyte*!
DISTR. **K**4; widely cultivated, but of unknown origin.
HAB. Waste and abandoned cultivated ground; ± 1950 m.

NOTE. The well-known flax cultivated for fibre and oily seeds, occasionally escaping and becoming naturalized. Easily recognized from the native species because of the normally blue flowers and large capsules.

2. **L. volkensii** *Engl.* in P.O.A. C: 226 (1895); Lind & Tallantire, Fl. Pl. Uganda: 62 (1962). Type: Tanganyika, Kilimanjaro, *Volkens* 369 (B, holo.†, K, iso.!)

Erect herb up to 90 cm. high, annual or perennial, often woody at the base, glabrous in all parts. Stems terete, obscurely longitudinally ridged. Leaves all alternate, linear-lanceolate to linear, 5–25 mm. long, 1–3 mm. wide, apiculate or sharply acuminate, margin scabrid-setulose, rarely entire. Stipules globose, small, dark brown, persistent. Inflorescence a many-flowered, lax or compact, monochasial, paniculate or corymbose cyme; pedicels 1–11 mm. long; bracts linear. Sepals 5, ovate or elliptic-ovate, long-acuminate, with glandular-serrulate margin; midrib prominent only in basal half to two-thirds of sepal. Petals 5, obovate, rounded or bluntly apiculate, yellow or orange-yellow, often fading white before falling, also occasionally purple-veined. Stamens 5. Styles 5; stigmas capitate. Capsule globose, 2–3 mm. in diameter, equalling or shorter than the persistent calyx, straw-coloured. Seeds flattened, ± 1 mm. long, shining, light brown. Fig. 3.

UGANDA. Toro District: Nyakasura, 7 Jan. 1936, *Hancock* 109/36!; Kigezi District: Kinaba Gap, Dec. 1938, *Chandler & Hancock* 2636!; Mbale District: Kapchorwa, Sebei, 7 Sept. 1954, *Lind* 221!
KENYA. Northern Frontier District: Mt. Nyiro, 30 Dec. 1955, *Adamson* 547!; Naivasha District: S. Kinangop, Kibata, 24 Nov. 1959, *Kerfoot* 1473!; Machakos District: Chumbi Hill, 30 June 1963, *Verdcourt* 3672!
TANGANYIKA. Mbulu District: Mbulumbul, salt-pan area, 28 June 1945, *Greenway* 7496!; Iringa District: Mufindi, near Irundi Hill, Sept. 1959, *Procter* 1406!; Songea District: Matengo Hills, 2·5 km. E. of Ndengo, 4 Mar. 1956, *Milne-Redhead & Taylor* 8975!
DISTR. **U**1–3; **K**1–6; **T**2–4, 7, 8; S. Ethiopia, E. Congo Republic, Malawi, Mozambique.
HAB. Upland grassland, in marshes and by streams; 1300–2750 m.

SYN. *L. gallicum* L. var. *holstii* Engl. in Phys. Abhandl. K. Akad. Wiss. Berl. 1894(1): 58 (1894), *nomen nudum*
  [*L. gallicum* L. var. *abyssinicum* sensu auct., e.g. Engl. in E.J. 30: 336 (1901), *non* Planch.]
  [*L. abyssinicum* sensu Robyns, F.P.N.A. 1: 404, t. 40 (1948), *non L. gallicum* L. var. *abyssinicum* Planch.]
  L. holstii Wilczek in B.J.B.B. 25: 311 (1955); F.C.B. 7: 38 (1958); F.Z. 2: 97 t. 10/B (1963). Type: Tanganyika, Usambara Mts., *Holst* 8988 (K, holo.!)

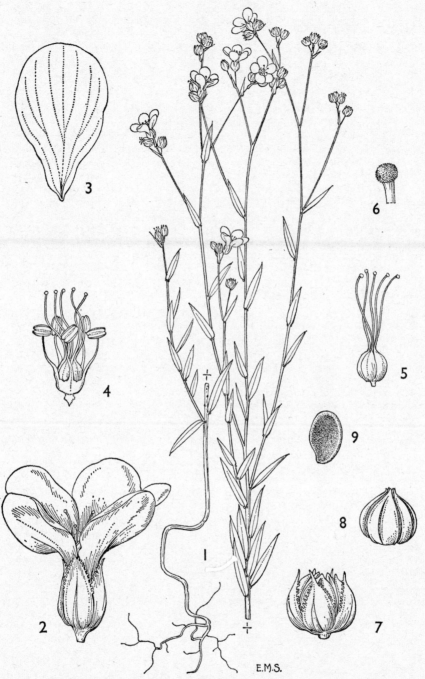

E.M.S.

FIG. 3. *LINUM VOLKENSII*—**1**, habit, × 1; **2**, flower, × 6; **3**, petal, × 6; **4**, stamens and gynoecium, × 6; **5**, gynoecium, × 6; **6**, stigma, × 60; **7**, fruit and persistent calyx, × 6; **8**, capsule showing septicidal and loculicidal planes of dehiscence, × 6; **9**, seed, × 12. 1, from *Stolz* 232; 2–9, from *Verdcourt* 3672.

VARIATION. As might be expected, specimens vary with geographical distribution. Forms differ in the degree of condensation of the inflorescence; there are both very lax- and compact-flowered specimens; the latter seem the most common.

3. L. thunbergii *Eckl. & Zeyh.*, Enum. Pl. Afr. Austr. Extratrop.: 35 (1834–5), excl. syn.; C.F.A. 1: 242 (1951); F.Z. 2: 97, t. 10/A (1963). Types: 3 syntypes from South Africa, Cape Province, one at Kew, without locality or collector (K, isosyn.!)

Erect or semiprocumbent herb up to 60 cm. high, perennial but may flower in first year. Stem terete, obscurely longitudinally ridged, nearly always glabrous. Leaves opposite or whorled on lower part of stem, becoming alternate towards the inflorescence, variable in shape often on one plant, from broadly elliptic-ovate or lanceolate to linear, 10–25 mm. long, 1–5 mm. wide, entire, acute or mucronate. Stipules globose, small, brown, or absent. Inflorescence a many-flowered, lax or compact, monochasial or dichasial paniculate or corymbose cyme; bracts linear; pedicels 1–5 mm. long. Sepals 5, elliptic or ovate-elliptic, 2–4 mm. long, 1·5–2 mm. broad, with capitate-glandular denticulate margin; midrib prominent, extending to the acuminate or apiculate apex. Petals 5, obovate, usually retuse, yellow, pale yellow or rarely white. Stamens 5. Styles 5; stigmas capitate. Capsule globose, 2·5–3 mm. in diameter, as long as or just exceeding the persistent calyx when mature, light brown. Seeds flattened, 1–1·2 mm. long, smooth and shining, light brown.

TANGANYIKA. Lushoto District: W. Usambara Mts., escarpment near Gologolo-Mkumbala footpath, 4 June 1953, *Drummond & Hemsley* 2855!; Mpanda District: Mahali Mts., E. face of Sisaga, 28 Aug. 1958, *Mgaza in Jefford, Juniper & Newbould* 1850!; Mbeya Mt., 13 May 1956, *Milne-Redhead & Taylor* 10325!
DISTR. **T**3, 4, 7; Malawi, Zambia, Rhodesia, Angola, and South Africa.
HAB. Upland grassland, in marshes, among rocks and by streams; 1750–2700 m.

4. L. keniense *T.C.E.Fries* in N.B.G.B. 8: 555, fig. 2 (1923). Type: Kenya, W. Mt. Kenya, *Fries* 1244 (UPS, holo., K, iso.!)

Perennial herb, usually prostrate or decumbent, up to 15–60 cm. high. Stem angular when young, becoming terete, leafless and ± 1 mm. in diameter when older; all parts glabrous. Leaves in whorls of 3 or 4, or opposite, rotund, broadly elliptic or broadly ovate, 9–13 mm. long, 4–7 mm. wide, always rounded at apex, entire. Stipules papillose, minute, greenish or absent. Inflorescence a few-flowered dichasial cyme; bracts leaf-like; pedicels 0·5–2 mm. long. Sepals 4, unequal in size, leaf-like, elliptic or elliptic-obovate, 4–7 mm. long, rounded at apex. Petals 4, obovate, 4–6·5 mm. long, shortly apiculate, bright yellow. Stamens 4. Styles 4, free; stigmas capitate. Capsule globose, 2–2·5 mm. in diameter, shorter than the persistent sepals and enclosed by them, straw-coloured. Seeds flattened, ± 1·2 mm. long, smooth, honey-brown.

UGANDA. Kigezi District: Behungi, 23 Dec. 1933, *A. S. Thomas* 1195!
KENYA. Elgon, May 1948, *Hedberg* 142!
TANGANYIKA. Kilimanjaro, just below Bismarck Hut, 24 Feb. 1953, *G. H. S. Wood* 934!; Rungwe District: E. side of Rungwe crater, Oct. 1959, *Procter* 1449!; Njombe District: Ndumbi R., 19 Oct. 1956, *Richards* 6624!
DISTR. **U**2; **K**3, 4; **T**2, 7; Ethiopia
HAB. Upland grassland, often near streams; also in clearings of bamboo zone; 2200–3250 m.

SYN. *L. keniense* T.C.E.Fries var. *aberdaricum* T.C.E. Fries. in N.B.G.B. 8: 556 (1923). Type: Kenya, Mt. Aberdare, *Fries* 2364 (UPS, holo., K, iso.!)
    [*L. gallicum* sensu W.F.K.: 21 (1948), *non* L.]

NOTE. One of the few tetramerous species of *Linum*.

3. RADIOLA

Hill, Brit. Herbal: 227 (1756); Roth, Tent. Fl. Germ. 1: 71 (1788)

Annual herb, glabrous in all parts. Leaves sessile, simple, opposite, entire. Stipules absent. Inflorescence a dichasial corymbose cyme. Sepals 4, toothed at apex, united at base, persistent. Petals 4. Stamens 4; rarely 4 staminodes also present; filaments united at base with petals. Ovary 4-locular; each loculus partly divided by a false septum, 2-ovulate; styles 4, free; stigmas capitate. Fruit a 4-locular, 8-valved, lobed capsule. Seeds small, ovoid, with little endosperm; embryo straight.

A widely distributed monotypic genus, recorded only once so far from the area covered by the Flora, namely from southern Tanganyika. It may be present in other areas of similar altitude, but is easily overlooked.

R. linoïdes *Roth*, Tent. Fl. Germ. 1: 71 (1788); Hegi, Ill. Fl. Mittel-Europa 5(1): 2, fig. 1662 (1924); F.W.T.A., ed. 2, 1: 361 (1958); Clapham, Tutin & Warburg, Fl. Brit. Is., ed. 2: 300 (1962); F.Z. 2: 99, t. 10/C (1963). Type: probably from Europe, *Herb. Linnaeus* 396.38 (LINN, syn.)

Minute erect herb, 1–7 cm. high at flowering. Leaves obovate, ovate or elliptic, 1·5–3 mm. long, 1–2 mm. wide, acute or rounded. Flowers terminal, in a usually repeatedly dichotomous dichasial cyme. Sepals (2–)3(–4)-toothed at apex, up to 1 mm. long. Petals obovate, as long as sepals, rounded or obtuse-apiculate, broadly clawed, white. Stamens as long as petals. Ovary 8-lobed; styles short. Capsule nearly as long as sepals, 8-seeded. Seeds ± 3 mm. long, smooth and shining, light brown. Fig. 2/5, p. 5.

TANGANYIKA. Njombe District: Poroto Mts., between Kikondo and Matamba, 17 May 1957, *Richards*!
DISTR. **T7**; scattered in mountainous areas of Africa—Cameroon Mt., Ethiopia, Malawi (Nyika Plateau); also in Madeira, Tenerife, N. Africa, and Europe from the Mediterranean north to Scandinavia
HAB. Upland grassland, among grass tussocks; ± 2300 m.
SYN. *Linum radiola* L., Sp. Pl.: 281 (1753). Type: as for *Radiola linoïdes* Roth

INDEX TO LINACEAE